PUBLIC LIBRARY DISTRICT OF COLUMBIA

LIGHT AND COLOR

Published in the United States in 2010 by
Stargazer Books, distributed by
Black Rabbit Books
P.O. Box 3263
Mankato, MN 56002

Illustrator: Tony Kenyon

Printed in the United States

Library of Congress Cataloging-in-Publication Data

Gibson, Gary, 1957-
  Light and color / Gary Gibson.
      p. cm. -- (Fun science projects)
  Includes index.
  ISBN 978-1-59604-188-2
 1. Light--Juvenile literature. 2. Light--Experiments--Juvenile literature.
3. Color--Juvenile literature. 4. Color--Experiments--Juvenile literature.
I. Title.
  QC360.G532 2009
  535--dc22

                              2008016395

Fun Science Projects
LIGHT AND COLOR

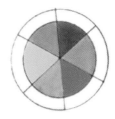

GARY GIBSON

Stargazer Books
Mankato, Minnesota

CONTENTS

INTRODUCTION

Have you ever wondered what would happen to the earth if there were no sunlight? Did you know that light rays can bounce, bend, and even be split into a rainbow? For centuries scientists worked to find out more about light and color. This book contains a selection of exciting "hands-on" projects to help explain some of the fascinating discoveries that have been made about light and color.

GET AN ADULT TO DO THIS FOR YOU

When this symbol appears, adult supervision is required.

LIGHT FOR LIFE

Green plants need sunlight to live and grow. They use the light's energy to grow. All animals get their food from plants, either directly or indirectly. Since plants need sunlight to grow, all living things depend on the sun.

GROWING WATERCRESS

1 Put a layer of cotton in the bottom of two clean dishes. Add a little water. Sprinkle watercress seeds evenly over the cotton.

2 Put the dishes on a sunny windowsill and cover each dish with a cardboard box. Make a hole in the side of one box and leave for several days. Check daily that the cotton is still damp.

3 The seeds under the box with no hole have grown straight up looking for light. The watercress under the box with the hole has grown toward the light.

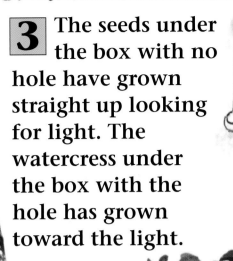

WHY IT WORKS

Green plants contain a chemical called chlorophyll. Chlorophyll traps light, which combines with water and air to help make plants grow. This process is called "photosynthesis." Plants cannot see light but can bend and grow toward where it comes from.

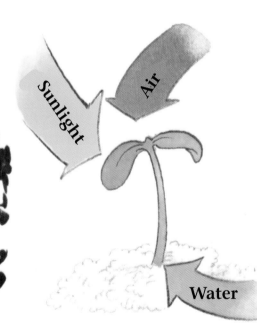

Sunlight

Air

Water

FURTHER IDEAS

Follow step 1 again. Cover one dish with a large, clean glass jar to make a "greenhouse." Compare the growing roots and shoots with the uncovered dish. Which seedlings grow the best?

DAY AND NIGHT

Half the world is in daylight and the other half in darkness. As the earth spins around, each part takes its turn to face the sun. Parts of the earth facing away from the sun can be lit only by the moon. Sometimes the moon passes between the sun and the earth, so the sun's rays are blocked and the sky grows dark. This is called an eclipse.

MAKE A SUNDIAL

1 You need a piece of wood, or thick cardboard, and a length of dowel. Make a hole near one edge of the wood for the dowel.

2 Stand the dowel in the hole (fix with glue if necessary). Decorate using waterproof paints.

3 On a sunny morning, put the sundial outside. The dowel casts a shadow; paint along the shadow.

4 Repeat step 3 every hour. Paint the time next to each shadow. The sundial will only work on sunny days. Remember to keep it in the same place, facing the same way.

WHY IT WORKS

The stick blocks the sun and casts a shadow. The shadow's position changes as the sun moves across the sky.

As the earth spins around, the sun appears to move across the sky.

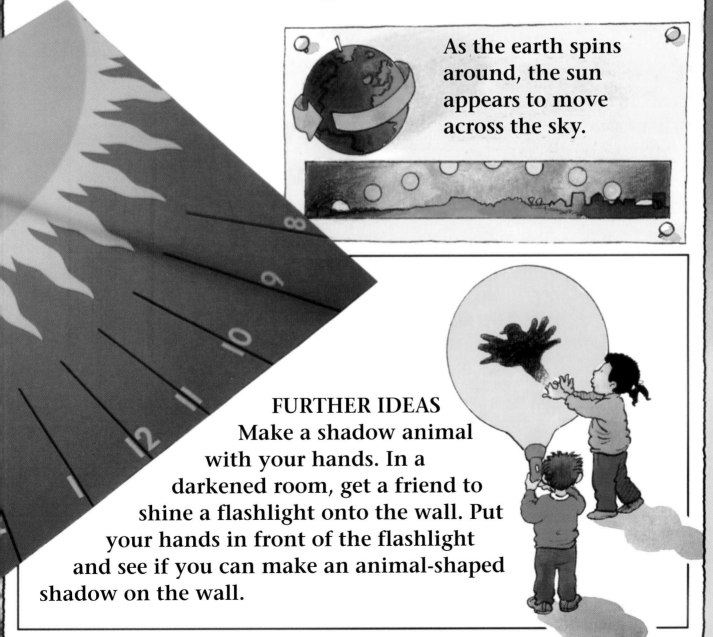

FURTHER IDEAS
Make a shadow animal with your hands. In a darkened room, get a friend to shine a flashlight onto the wall. Put your hands in front of the flashlight and see if you can make an animal-shaped shadow on the wall.

SEEING IMAGES

An image is a likeness of something or someone. What we see in photographs or a movie are images. A camera is a box that can make an image on photographic film. The film contains chemicals that will keep the image for years.

MAKE A PINHOLE CAMERA

1 Find a small cardboard box that does not let light through. Use a pair of scissors to cut out one side of the box.

2 Tape a piece of tracing paper over the cutout side of the box. Make sure that the tracing paper is kept as smooth as possible.

3 Make a pinhole in the side of the box opposite the tracing paper. Point the pinhole at the window until you see its image on the tracing paper.

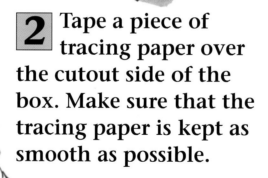

WHY IT WORKS

Light rays in air do not bend or curve; they always travel in straight lines. Rays of light from the window enter the pinhole in straight lines and hit the tracing paper. The light rays from the bottom and the top of the window cross over as they pass through the hole, so the image appears upside down.

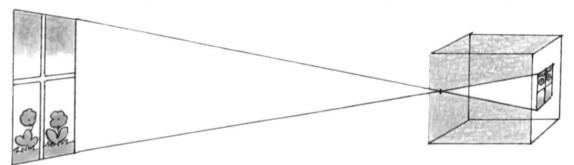

FURTHER IDEAS
Make the pinhole a little bigger so more light enters the camera. The image becomes brighter but less clear. A magnifying glass in front of the pinhole can sharpen the image. The image will be faint, so point the pinhole at a bright object such as a lightbulb.

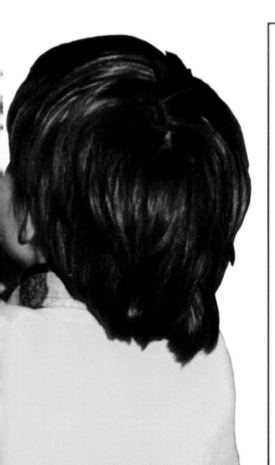

BOUNCING LIGHT

Rays of light can be bounced off an object like a rubber ball bouncing off a wall. We call this "reflection." Light rays are reflected best by flat, shiny surfaces such as shiny spoons, cans, bottles, or mirrors.

MAKE A KALEIDOSCOPE

1 Carefully tape together three identical-sized small mirrors. Make a triangular tube with the shiny sides facing inside.

2 Cut out a triangle-shaped piece of paper, allowing for flaps. Tape it over one end of the triangle of mirrors to form a box.

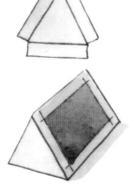

3 Cut out small pieces of brightly colored paper from a magazine and drop them into the bottom of the box.

GET AN ADULT TO HELP YOU

4 Tape another triangle of paper over the other end of the tube. Using a pencil, make a hole to look through. The kaleidoscope is finished.

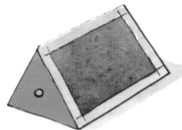

5 Hold the kaleidoscope level, and point it at a bright light. Look at the pattern through the eye piece. Shake and look again.

WHY IT WORKS

Light rays from the colored paper are reflected back and forth between the mirrors. Each image is doubled by the mirrors before the light rays reach your eye.

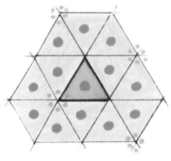

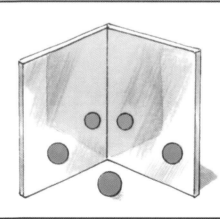

FURTHER IDEAS
Stand two mirrors up at right angles, using modeling clay. Place a marble between the mirrors. How many images can you see?

UP PERISCOPE!

Submarine crews want to know what is going on above the waves without being seen. Instead of rising to the ocean's surface, the submarine raises its periscope. On land you can use periscopes to see over walls and around corners!

MAKE A PERISCOPE

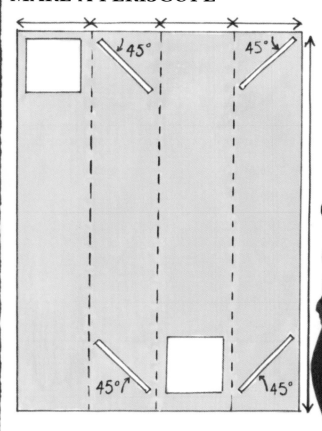

45°
45°
45°
45°

1 Copy this pattern onto cardboard. Cut around the outline. Cut out the slots and squares. Fold on the dotted lines.

2 Tape the edges of the cardboard together to form a box. Make sure that the slots line up. Paint to decorate.

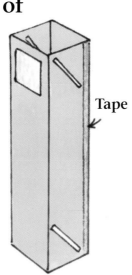

Tape

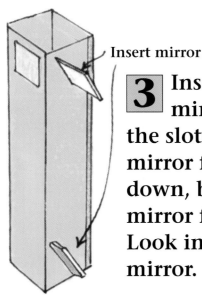

Insert mirror

3 Insert two mirrors into the slots, top mirror facing down, bottom mirror facing up. Look into the lower mirror.

WHY IT WORKS

Light rays above and ahead of you hit the upper mirror. It reflects the rays down to the lower mirror, which in turn reflects the light rays into your eyes.

FURTHER IDEAS
Make a periscope to see around corners. Copy the design shown below. Follow instructions 1 and 2 as before. But this time the angles of the mirror slots are different. Make sure the top mirror faces down and the bottom mirror faces up.

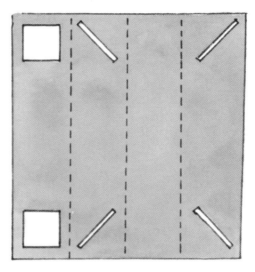

MOVING PICTURES

Cartoon films are made out of many drawings. These are photographed one after another by a movie camera. When the film is projected onto a screen, the images seem to move. If you move your eyes quickly over the pictures on the right, the ball seems to bounce.

MAKE A FLICK BOOK

1 Draw a background picture. Trace it onto at least 12 pages of the same size. Leave a margin down one side of each drawing.

2 Draw the sun high in the sky on the first page. Draw it slightly lower on the second page. Repeat until the sun has set on the last page.

3 Stack the pages neatly and staple them together with two staples along the edge of the margin.

WHY IT WORKS

Your eyes see each image for a fraction of a second. If the images are shown fast enough, the eye runs the images together. Differences in the separate images appear as movement.

FURTHER IDEAS
Copy the two faces (below) onto two sheets of tracing paper. Staple the two sheets together. Roll the upper sheet tightly around a pencil. Move the pencil up and down to roll and unroll the upper paper.

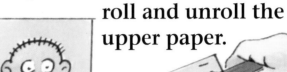

4 Hold the flick book by the margin and watch the sun go down as you flick the pages.

SPLITTING LIGHT

More than 300 years ago, Sir Isaac Newton proved that white light is made from the colors of the rainbow. Newton split white light into a rainbow using a wedge of glass called a prism. We see rainbows in the sky because water droplets in the air split the sunlight before it reaches us.

2 Angle a mirror in a bowl of water. Bend a large piece of white posterboard away from the bowl.

MAKE A
RAINBOW

1 Get an adult to cut a slit in a piece of black posterboard. Shine a lamp through the slit to be sure you get a narrow beam of light.

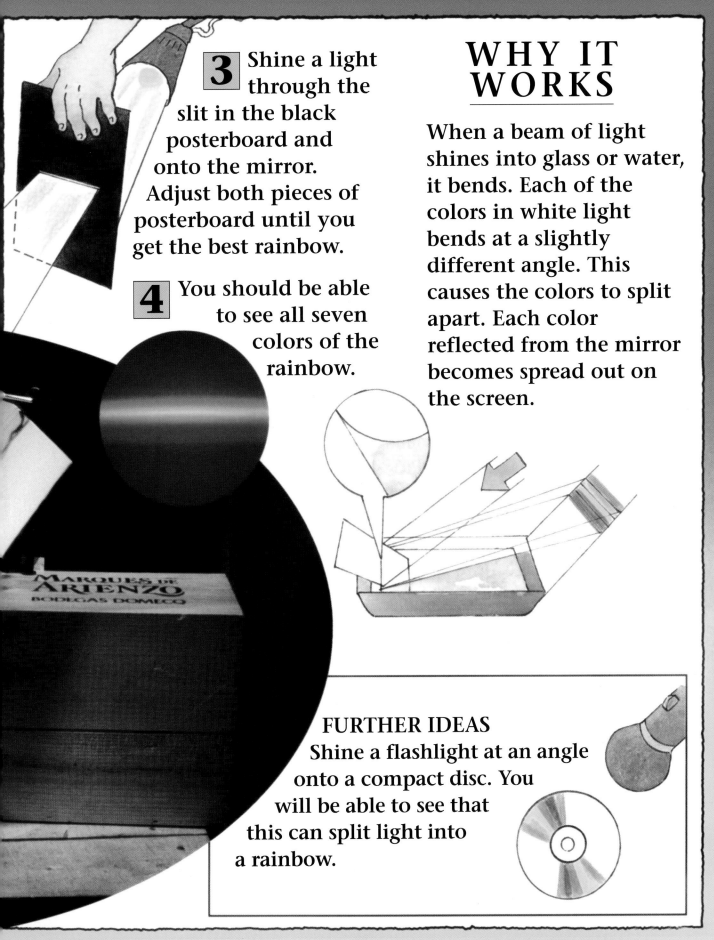

3 Shine a light through the slit in the black posterboard and onto the mirror. Adjust both pieces of posterboard until you get the best rainbow.

4 You should be able to see all seven colors of the rainbow.

WHY IT WORKS

When a beam of light shines into glass or water, it bends. Each of the colors in white light bends at a slightly different angle. This causes the colors to split apart. Each color reflected from the mirror becomes spread out on the screen.

FURTHER IDEAS
Shine a flashlight at an angle onto a compact disc. You will be able to see that this can split light into a rainbow.

MIXING COLORS

Look closely at a color TV or the photographs in this book. The pictures are made up of lots of tiny, colored dots.

Because we see books or TV from a distance, the dots seem to mix to make colors.

MAKE A COLORED SPINNER

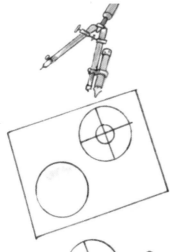

1 Use a pencil and a pair of compasses to draw circles of different sizes onto white posterboard. Cut them out with scissors.

2 Divide the circles into equal sections and decorate each section with different colors. Push a sharp pencil or stick through the hole in the center of each circle.

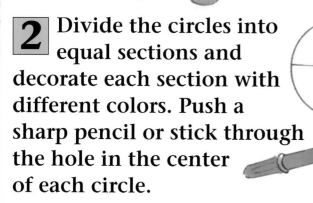

3 Spin the spinner as fast as you can on a tabletop and watch the different colors merge. If you color a spinner with the colors of the rainbow, it may appear white when you spin it.

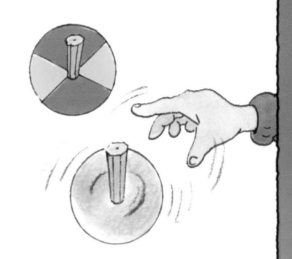

WHY IT WORKS

The spinner is turning so fast that instead of seeing separate colors, our eyes see a mixture. White light is made up of the colors of the rainbow, so a spinner decorated with these colors appears white.

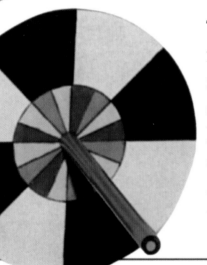

FURTHER IDEAS
Cover three flashlights with red, blue, and green cellophane. Shine them onto white paper (or a white floor). Allow the light beams to overlap. See how many new colors you can make.

SEEING IN THREE-DIMENSION

Animals usually have two eyes. Close one eye and look at an object. Guess how far away it is. Try again with both eyes open. It is much harder to judge distances using only one eye. Having two eyes gives us a sense of depth.

MAKE 3-D GLASSES

1 Measure the distances A and B around your head with a tape measure.

2 Use the distances to draw out your glasses onto cardboard. Cut out the glasses and fold along the dotted lines.

3 Cut out red and green cellophane for the eyeholes. Glue the green over the right eyehole and red over the left. Try on the glasses. Look at the insect picture on page 23.

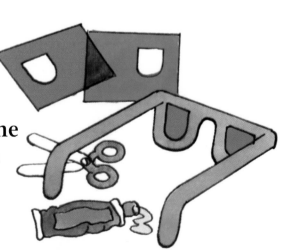

WHY IT WORKS

The left eye covered with red cellophane only sees the green image; the right eye only sees the red. Each eye sees a different view. The brain combines both images to give the illusion of depth.

FURTHER IDEAS

Close one eye. Hold your hands out in front of you at arm's length. Bring your index fingers together so they touch. Can you thread a needle with one eye closed?

SEPARATING COLORS

There are three primary colors in printing and painting—red, blue, and yellow. The enormous range of colored dyes, paints, and inks are made by mixing different amounts of two or more of the primary colors.

FIND THE HIDDEN COLORS

1 With a pair of compasses draw some circles onto paper towels. Cut them out with scissors.

2 Using marker pens, draw a dot of color in the middle of each circle. Black, purple, green, brown, and orange are good colors to use.

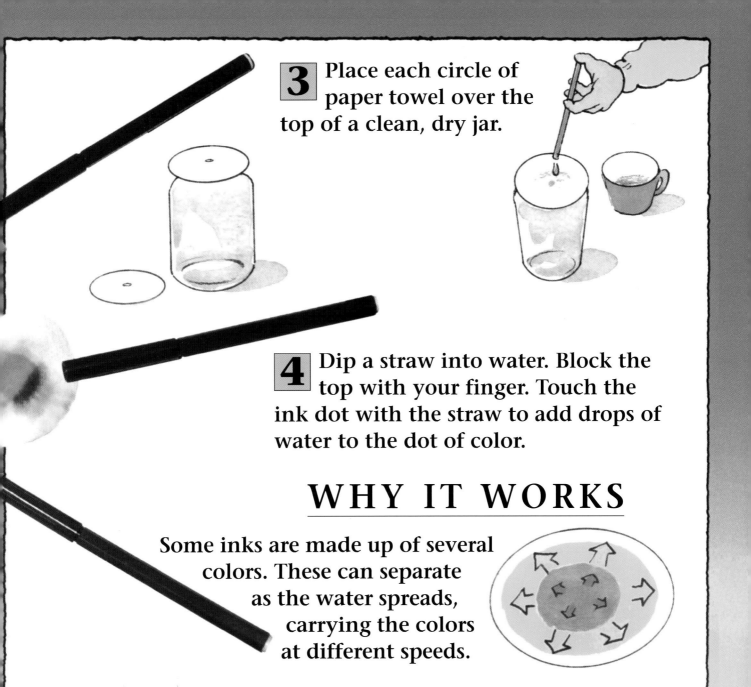

3 Place each circle of paper towel over the top of a clean, dry jar.

4 Dip a straw into water. Block the top with your finger. Touch the ink dot with the straw to add drops of water to the dot of color.

WHY IT WORKS

Some inks are made up of several colors. These can separate as the water spreads, carrying the colors at different speeds.

FURTHER IDEAS
Take a long strip of paper towel. Draw a large dot near the bottom. Hang the strip up so the end just dips into a bowl of water. Watch the colors separate as the water rises up the paper.

COLORED DYES

Today we can buy clothes in an enormous variety of colors. These colors come from modern artificial dyes made from oil. Before the last century, people had always used natural dyes made from plants, animals, or materials in the ground.

TIE-DYE A HANDKERCHIEF

GET AN ADULT TO DO THIS FOR YOU

2 Tie some string around a white cotton handkerchief as tightly as you can.

1 Collect lots of brown onion skins. Ask an adult to boil them in water for 20 minutes.

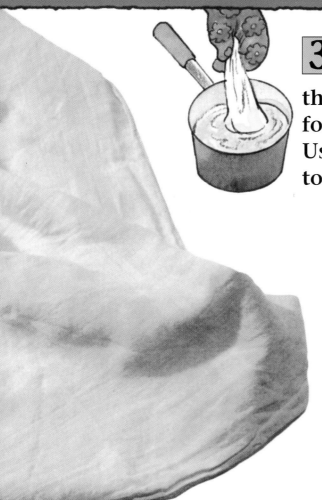

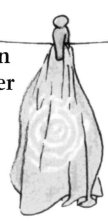

3 Soak the tied handkerchief in the onion skin water for five minutes. Use an oven glove to protect yourself.

4 Cut the string from the dyed handkerchief when cool. Fasten the handkerchief to a clothesline until it is dry.

WHY IT WORKS

Onion skins contain a chemical called a pigment. Boiling brings out the pigment, which in onion skin is yellow. Compare how well the pigment dyes fabrics other than cotton.

FURTHER IDEAS
Many vegetables contain different colored pigments. See what color beet juice or spinach water dye fabric.

COLOR CHANGES

Lemons taste sour because they contain acid ("acid" means sour). Hundreds of chemicals are acid. It would be very dangerous if scientists had to taste chemicals to identify them. Instead they use a chemical that changes color when acid is added.

GET AN ADULT TO DO THIS FOR YOU

TESTING FOR ACIDS

1 Ask an adult to chop up half a red cabbage. Boil the chopped cabbage in a pan of water for about five minutes.

2 Remove the cabbage from the water. Cut paper towels or filter paper into strips.

3 Dip each strip into the cabbage water. Allow the strips to soak the water up.

4 Let the strips dry. When dry, try adding drops of vinegar, lemon, soap, and other harmless substances to each strip.

WHY IT WORKS

Red cabbage contains a chemical called an indicator. Indicators change color when an acid or alkali is added. Red cabbage juice turns red in acids and green in alkalis. Litmus is an important indicator commonly used by scientists.

5 Note the different colors you see on each strip.

FURTHER IDEAS
You can use geranium petals instead of red cabbage. Geranium petals also contain an indicator that changes color when an acid or alkali is added.

FANTASTIC LIGHT FACTS

The sun is 93,000,000 miles (150 million km) from Earth. But because light travels at an amazing 186,000 miles per second (300,000,000 meters per second), it takes only eight minutes for the sun's light to reach us.

The ancient Greeks believed that the sun was driven from east to west in a flaming chariot, guided by the sun god Helios.

A peacock's bright feathers are created by light reflecting off microscopic fringes, called barbules, on each feather.

The world's first photograph was taken in 1826 by a French inventor named Joseph Nicéphore Niepce. He captured the view through his window. But the picture took eight hours to take!

The brightest artificial light was produced in March 1987 by a laser at the Los Alamos National Laboratory in New Mexico. It was several million times brighter than the sun.

The longest lasting lightbulb can be found in Livermore, California. It was first switched on in 1901.

In a desert, people sometimes think they can see objects or a pool of water in the distance. This is really an image called a mirage.

Mirages happen when air near the ground is much hotter than air higher up. As light from the sun passes from cooler to warmer air, it travels faster and is bent upward so that images appear where there is nothing.

Inside our eyes we have light-sensitive cells called cones. These recognize different colored light. The eye of someone with normal color vision has three types of cones. One type recognizes red light, one blue, and the third green light.

A color-blind person will have one or more types of cones missing. To them, certain colors such as red and green will look the same.

GLOSSARY

Acid
A liquid that turns blue litmus (an indicator) red.

Alkali
A liquid that turns red litmus (an indicator) blue.

Chlorophyll
The chemical pigment that gives green plants their color. It traps the energy contained in sunlight needed for photosynthesis.

Image
The "picture" of an object usually formed by a lens or photograph.

Indicator
A chemical that changes color when an acid or alkali is added.

Light ray
A very narrow beam of light.

Litmus
An indicator that turns red in acids and blue in alkalis.

Photosynthesis
A chemical process where light energy trapped by chlorophyll combines with water and air to help make a plant grow.

Pigment
The substance added to paints and dyes to give them color.

Primary colors
There are three primary colors of paints and dyes, from which all other colors are made: red, yellow, and blue.

Prism
A transparent wedge, usually of glass, used to split white light into the colors of the rainbow.

Reflect
When light or sound is bounced back from a surface.

Shadow
A place of darkness created by an object blocking light.

INDEX